The Scientist's Thought Catalogue

Copyright © 2016

All rights reserved.

ISBN-13: 978-1537443249
ISBN-10: 1537443240

www.ingramcontent.com/pod-product-compliance
Lightning Source LLC
Chambersburg PA
CBHW070227190526
45169CB00001B/111